My Book

This book belongs to

Name:_____

Copy right © 2019 MATH-KNOTS LLC

All rights reserved, no part of this publication may be reproduced, stored in any system or transmitted in any form, or by any means, electronic, mechanical, photocopying, recording, or otherwise without the written permission of MATH-KNOTS LLC.

Cover Design by :
Gowri Vemuri

First Edition :
April, 2020

Author :
Gowri Vemuri

Editor :
Ritvik Pothapragada

Questions: mathknots.help@gmail.com

Dedication

This book is dedicated to:

My Mom, who is my best critic, guide and supporter.

To what I am today, and what I am going to become tomorrow,

is all because of your blessings, unconditional affection and support.

This book is dedicated to the

strongest women of my life,

my dearest mom

and

to all those moms in this universe.

G.V.

QUANT - Q

INDEX

Notes	9 - 10
Decimal : Additions	11 - 27
Decimal : Subtraction	28 - 42
Decimal : Addition & Subtraction	43 - 51
Decimal : Multiplication	52 - 66
Decimal : Division	67 - 76
Rounding decimals	77 - 86
Write the numeral in words	87 - 95
Write each word as numeral	96 - 119
Write the decimal place underlined	120 - 134
Rounding the decimal	135 - 144
Answer Keys	145 - 178

DECIMALS — Notes

Decimals notes

The word standard means regular. Numbers in standard form are whole numbers or natural numbers.

Example : The number "six hundred twenty five" in standard form is 625.

To name a decimal from its standard form, follow these steps :

1. Name the number in front of the decimal. (Do not include the word "and").

2. The word "and" is used for the decimal point.

3. The number in the decimal part is similar to the number in front of the decimal.

4. Name the last place value given (of the digit farther to the right).

Rounding Decimals to a Given Place Value:

To round a decimal number to a given place value, look at the digit to the right of the desired place value and follow the rounding rules:

"5 and above, give it a shove! 4 or below, leave it alone!"

(The number in the desired place value gets "bumped up" to the next consecutive value if the digit to the right of it is 5 or more. The number in the desired place value does not change if the digit to the right of it has a value of 5 or less.)

The purpose of rounding is to provide *an* estimate and get an approximate value. Rounding involves losing some accuracy.

Example : If 5,953 people attend a Soccer game. We can approximately say 6,000 people watched the game.

Comparing and Ordering Decimals:

To order the given decimals, compare the digits of all decimals according to their place value.

Tip : Line the numbers up vertically according to their place value and compare from the left to the right. When comparing like place values from the left, the number with the higher digit is the larger number.

Also, added zeros to the right of a decimal number does not change the value of the decimal.

For example, 0.8 = 0.80 = 0.800 = 0.8000 = 0.80000

DECIMALS

Naming/Reading Decimal :

Read the number to the left of the decimal. Say the word "and" for the decimal place. Read the number to the right of the decimal as you would read it if the number were on the left. End the name with the place value of the digit that is furthest to the right.

Converting Between Fractions and Decimals :

Fractions and decimals represent a certain "part of a whole", all fractions can be written as decimals and terminating/repeating decimals can be written as fractions.

Multiplying Decimals :

To multiply decimals, first ignore the decimals and simply multiply the digits. Then count the total number of spaces from the right to the decimal (or digits to the right of the decimal) of both numbers; place the decimal that number of spaces from the right into your answer.

Dividing Decimals:

Divide the decimals as a whole numbers ignoring the decimals.

1. Count the number of digits of the dividend after the decimal.
2. Count the number of digits of the divisor.
3. Subtract the number obtained in step 1 from the number obtained instep 2.
4. Count the number of digits from the right to the number obtained in step 3.
5. Place the decimal point.

DECIMALS

Basic Math

Evaluate the below decimals

(1) 55.137 + 64.71

(2) 16.2 + 91.8

(3) 19.8 + 97.5

(4) 42.4 + 78.58

(5) 6.5 + 79.3

(6) 94.7 + 88.5

(7) 39.9 + 63.4

(8) 33.6 + 74.8

(9) 95.8 + 7.2

(10) 24.2 + 30.5

DECIMALS

Evaluate the below decimals

(11) 41.4 + 41.3

(12) 36.42 + 0.7

(13) 33 + 68.35

(14) 64.17 + 38.7

(15) 71.1 + 72.8

(16) 96.5 + 62.1

(17) 0.3 + 14.4

(18) 18.2 + 60.9

(19) 8.2 + 77.4

(20) 68 + 92.3

DECIMALS

Evaluate the below decimals

(21) 51.7 + 43.6

(22) 93 + 70.7

(23) 9.81 + 36.06

(24) 32.5 + 77.6

(25) 86.8 + 84.37

(26) 1.8 + 95.2

(27) 84.7 + 99.9

(28) 4.66 + 6.5

(29) 3.4 + 73.1

(30) 33.87 + 40.9

DECIMALS

Evaluate the below decimals

(31) 7.3 + 88.2

(32) 97.1 + 37.1

(33) 40.9 + 18.2

(34) 62.6 + 39.082

(35) 57.8 + 81.8

(36) 58.9 + 85.8

(37) 64.89 + 19.239

(38) 33.7 + 66.3

(39) 51.9 + 49.7

(40) 75.5 + 86.7

DECIMALS

Evaluate the below decimals

(41) 56.8 + 39.7

(42) 25.9 + 98.4

(43) 38.3 + 6.6

(44) 21.74 + 83.479

(45) 17.28 + 70.8

(46) 71.6 + 44.68

(47) 2.214 + 78.1

(48) 71.8 + 87.84

(49) 56.4 + 17.6

(50) 44.7 + 23.8

DECIMALS

Evaluate the below decimals

(51) 234.6 + 481.7

(52) 525.3 + 34.7

(53) 427.5 + 140.7

(54) 300.173 + 969.8

(55) 709.3 + 151.6

(56) 550.3 + 196.3

(57) 313 + 873.1

(58) 871.482 + 412.5

(59) 617.3 + 696.3

(60) 50.9 + 273.8

DECIMALS

Evaluate the below decimals

(61) 39.2 + 803.7

(62) 501.2 + 823.8

(63) 316.1 + 723.11

(64) 142.5 + 372.4

(65) 791.9 + 179.8

(66) 532.6 + 860.1

(67) 533.63 + 872.3

(68) 161.6 + 237.2

(69) 94.9 + 923.8

(70) 820.7 + 830.4

DECIMALS

Evaluate the below decimals

(71) 772.3 + 319.9

(72) 531.1 + 354.078

(73) 411.8 + 346.7

(74) 98.8 + 412.2

(75) 140.7 + 761.3

(76) 612.7 + 683.11

(77) 198.8 + 127.8

(78) 678.8 + 356.4

(79) 524.9 + 119.1

(80) 261 + 203.4

DECIMALS

Evaluate the below decimals

(81) 542.8 + 943.1

(82) 618.1 + 843.4

(83) 659.4 + 27.5

(84) 434.4 + 119.95

(85) 188.9 + 62.5

(86) 142.9 + 499.8

(87) 953.5 + 401.4

(88) 348 + 297.6

(89) 510.3 + 925.9

(90) 168.5 + 509.7

DECIMALS

Basic Math

Evaluate the below decimals

(91) 883.047 + 964.9

(92) 747.5 + 876.8

(93) 823.8 + 243.809

(94) 150.69 + 271.07

(95) 157.6 + 399.5

(96) 118.8 + 798.1

(97) 213.5 + 844.06

(98) 819.5 + 393.6

(99) 963.1 + 378.9

(100) 246.5 + 383.7

DECIMALS

Evaluate the below decimals

(101) 690.01 + 554.7 + 256.4

(102) 884.6 + 287.6 + 86.4

(103) 934.7 + 370.3 + 492.13

(104) 701.1 + 200.6 + 277.9

(105) 954.5 + 948.9 + 722

(106) 465.6 + 94.7 + 343.2

(107) 902.6 + 584.7 + 886.5

(108) 194.4 + 571.8 + 518.1

(109) 411.9 + 224.6 + 712.5

(110) 940.5 + 538.21 + 674.7

DECIMALS

Evaluate the below decimals

(111) 722.1 + 40.3 + 274.39

(112) 350.8 + 635.67 + 86.3

(113) 413.2 + 429.9 + 697.9

(114) 567.7 + 32.2 + 746.6

(115) 128.9 + 984.9 + 482.2

(116) 642.231 + 58.894 + 164.8

(117) 523.7 + 96.1 + 568.2

(118) 747.97 + 192.74 + 763.5

(119) 572.2 + 24.5 + 215.7

(120) 186.06 + 713.7 + 509.6

DECIMALS

Basic Math

Evaluate the below decimals

(121) 313.9 + 16.7 + 496.9

(122) 520.498 + 290.8 + 476.4

(123) 810.3 + 424.1 + 347.1

(124) 600.9 + 667 + 444.7

(125) 781.7 + 234.4 + 860.9

(126) 287.4 + 742.1 + 431.34

(127) 602.6 + 724.8 + 123.4

(128) 718.2 + 528 + 922.02

(129) 838.5 + 319.7 + 453.73

(130) 147.7 + 71.41 + 766.4

DECIMALS

Basic Math

Evaluate the below decimals

(131) 270.228 + 687.5 + 712.55

(132) 59.9 + 477.16 + 74.3

(133) 928.1 + 326.6 + 513.1

(134) 186.06 + 717 + 904.5

(135) 210.4 + 330.1 + 231.7

(136) 279.178 + 170.1 + 228.8

(137) 692.3 + 997.5 + 123.442

(138) 244.9 + 800.4 + 356.8

(139) 356.2 + 886.7 + 568.3

(140) 596.628 + 756.5 + 308.2

DECIMALS

Basic Math

Evaluate the below decimals

(141) 392.8 + 973.1 + 621.7

(142) 847.2 + 903.3 + 952.66

(143) 969.48 + 941.98 + 903.4

(144) 260.8 + 166.5 + 458.5

(145) 522.04 + 904.2 + 423.1

(146) 201.7 + 609.329 + 323.4

(147) 308.3 + 945.6 + 106.7

148) 202.5 + 176.3 + 192.99

(149) 971.6 + 723.5 + 84.2

(150) 839.3 + 773.6 + 130.1

DECIMALS

Find the difference of the below decimals

(151) 63.7 − 43.15

(152) 90.5 − 61.614

(153) 92.518 − 53.5

(154) 80.05 − 69.4

(155) 36.3 − 33.13

(156) 20.8 − 17.6

(157) 73.3 − 2.19

(158) 93.1 − 11.7

(159) 84.8 − 79.3

(160) 68.5 − 13.2

DECIMALS

Basic Math

Find the difference of the below decimals

(161) 13.959 − 2.4

(162) 67.73 − 1.9

(163) 30.5 − 1.6

(164) 79.03 − 71.3

(165) 96.7 − 82.4

(166) 63.5 − 45.9

(167) 26.1 − 16.4

(168) 81.9 − 3.7

(169) 98.4 − 22.2

(170) 60.3 − 51.5

DECIMALS

Find the difference of the below decimals

(171) 87.2 − 73.354

(172) 18.3 − 7.7

(173) 82.3 − 25.96

(174) 84.46 − 73.7

(175) 46.4 − 24.3

(176) 83.4 − 62.1

(177) 94.55 − 58.1

(178) 87.1 − 52.8

(179) 94.9 − 17.8

(180) 49 − 25.7

DECIMALS

Find the difference of the below decimals

(181) 92.1 − 20.6

(182) 84.8 − 41.2

(183) 57.7 − 49.3

(184) 32.8 − 0.1

(185) 48.12 − 42.88

(186) 48.7 − 23.1

(187) 52.5 − 22.46

(188) 69.6 − 50.1

(189) 98.867 − 89.2

(190) 71.066 − 18.4

DECIMALS

Find the difference of the below decimals

(191) 82.6 − 69.8

(192) 62.5 − 27.62

(193) 21.8 − 8.69

(194) 40.3 − 12.4

(195) 96.3 − 61.6

(196) 29 − 13.4

(197) 98.3 − 46.59

(198) 50.2 − 17.5

(199) 90.56 − 16.7

(200) 66.8 − 8.1

DECIMALS

Basic Math

Find the difference of the below decimals

(201) 71.7 - 36.5

(202) 351.2 - 313.5

(203) 460.1 - 228.3

(204) 431.4 - 400.2

(205) 340.683 - 76.8

(206) 254.7 - 18.1

(207) 428.4 - 272.8

(208) 337.3 - 169.7

(209) 149.8 - 57.9

(210) 377.7 - 297.38

DECIMALS

Basic Math

Find the difference of the below decimals

(211) 195.5 − 163.5

(212) 229.9 − 68.4

(213) 476.6 − 454.4

(214) 457.8 − 261.9

(215) 474.9 − 435.9

(216) 450.9 − 332.6

(217) 90.86 − 25.7

(218) 346.65 − 242.2

(219) 402.7 − 56.4

(220) 209.56 − 29.688

DECIMALS

Find the difference of the below decimals

(221) 481.3 − 141.5

(222) 233.24 − 17.6

(223) 429.646 − 235.3

(224) 427.9 − 29.688

(225) 156.2 − 29.688

(226) 480 − 94.1

(227) 394.6 − 249.649

(228) 271.9 − 32.6

(229) 440.6 − 219.86

(230) 310.8 − 175.657

DECIMALS

Find the difference of the below decimals

(231) 357.7 − 43.04

(232) 353.9 − 4.48

(233) 456.7 − 434.6

(234) 243.2 − 162.9

(235) 65.53 − 53.7

(236) 292.2 − 88.8

(237) 411.5 − 100.9

(238) 138.4 − 80.4

(239) 449.7 − 191.38

(240) 435.4 − 220.94

DECIMALS

Basic Math

Find the difference of the below decimals

(241) 143.435 − 125.3

(242) 409.8 − 168.1

(243) 394.64 − 263.3

(244) 395.4 − 87.86

(245) 424.6 − 124.1

(246) 178 − 59.8

(247) 151.1 − 3.5

(248) 227.7 − 69.6

(249) 175.4 − 78.3

(250) 185.5 − 43.8

DECIMALS

Find the difference of the below decimals

(251)   58.2 − 20.3 − 15.726

(252)   43.7 − 15.943 − 0.9

(253)   98.8 − 7.5 − 51.62

(254)   14.2 − 10.6 − 0.4

(255)   79.9 − 12.5 − 8.4

(256)   63.6 − 51.3 − 7.8

(257)   76.7 − 20.6 − 40.1

(258)   96.3 − 18.8 − 14.3

(259)   96.7 − 52.1 − 19.2

(260)   42.3 − 30.4 − 4.37

DECIMALS

Basic Math

Find the difference of the below decimals

(261)   90.047 − 23.7 − 45.2

(262)   66.1 − 13 − 18.7

(263)   49.79 − 32.4 − 15.4

(264)   48.9 − 7 − 31.6

(265)   97.7 − 60.5 − 26.7

(266)   91 − 43.91 − 40.8

(267)   95.9 − 9.8 − 4

(268)   68.9 − 12.3 − 7.5

(269)   96.8 − 1.9 − 38.3

(270)   77.8 − 42.8 − 31.4

DECIMALS

Basic Math

Find the difference of the below decimals

(271) 82.4 − 34.916 − 16.8

(272) 88.9 − 10.5 − 69.2

(273) 81.6 − 38.3 − 27.1

(274) 91 − 2.3 − 42.4

(275) 78.1 − 20 − 4.8

(276) 96.2 − 35.9 − 50.4

(277) 91.3 − 35.32 − 39.6

(278) 90.79 − 1.4 − 27.302

(279) 97.9 − 67.8 − 14.5

(280) 97.07 − 8.6 − 72.9

DECIMALS

Basic Math

Find the difference of the below decimals

(281) 39.3 - 9.612 - 12.4

(282) 80.4 - 15.7 - 4.6

(283) 77.4 - 12.3 - 10.4

(284) 54.4 - 11.2 - 22.5

(285) 86.2 - 0.82 - 64.5

(286) 58.4 - 29.1 - 18.4

(287) 49.823 - 27.1 - 3.5

(288) 89.6 - 1.5 - 48.2

(289) 66.6 - 8.8 - 28.8

(290) 76.2 - 2.3 - 56.57

DECIMALS

Basic Math

Find the difference of the below decimals

(291)   73.2 − 13.5 − 29.5

(292)   75 − 16 − 6.99

(293)   83.5 − 50.8 − 11.4

(294)   96.1 − 23.2 − 44.8

(295)   55 − 19.1 − 26.6

(296)   88.4 − 50.09 − 27.7

(297)   82.2 − 44.9 − 0.7

(298)   69.19 − 21.7 − 14.1

(299)   98.9 − 10.9 − 38.5

(300)   83.1 − 21.9 − 20.07

DECIMALS

Evaluate each expression.

(301) 115.3 + 418.6

(302) 31.5 + 148.67

(303) 364.1 − 362.4

(304) 229 + 205.5

(305) 311.1 − 98.4

(306) 378 − 116.4

(307) 230.1 + 146.58

(308) 170.1 + 67.2

(309) 461.29 − 20.9

(310) 319.7 − 96.9

DECIMALS

Evaluate each expression.

(311)   173.2 + 142.8

(312)   132.47 + 182.2

(313)   468.1 + 14.4

(314)   443.8 + 94.6

(315)   74.6 + 410.2

(316)   393.5 − 17.7

(317)   357.91 − 120.5

(318)   57.686 + 159.9

(319)   370.8 + 348.4

(320)   271.2 + 358.5

DECIMALS

Evaluate each expression.

(321) 453.1 − 21.9

(322) 276.9 − 138.2

(323) 373.7 − 73.9

(324) 67.3 − 26.1

(325) 340.6 − 259.942

(326) 340.6 + 129.9

(327) 355.1 − 45.015

(328) 67.73 + 441.2

(329) 257.4 + 257.1

(330) 466.33 − 84.2

DECIMALS

Evaluate each expression.

(331) 203.5 + 47.5

(332) 196.5 − 143.8

(333) 69.9 + 186.889

(334) 161.68 + 292.6

(335) 236 + 280.6

(336) 463.2 − 427.3

(337) 173.1 + 412.3

(338) 65.5 + 428.2

(339) 144.641 + 165.26

(340) 429.3 − 194.6

DECIMALS

Basic Math

Evaluate each expression.

(341) 492.5 + 373.9

(342) 440.3 − 37.63

(343) 412.8 − 91.7

(344) 77.35 + 230.6

(345) 492.5 − 29.4

(346) 467.9 + 230.4

(347) 43 + 310.3

(348) 56.8 + 155.5

(349) 363.2 + 388.46

(350) 151.1 + 377.2

DECIMALS

Evaluate each expression.

(351) 384.02 − 131.7

(352) 321 − 209.7

(353) 119.2 + 202.9

(354) 364.9 + 99.8

(355) 402.6 − 30.1

(356) 272.09 − 132.1

(357) 163.4 + 81.8

(358) 255.9 + 210.5

(359) 350.4 − 172.84

(360) 77.5 + 377.94

DECIMALS

Evaluate each expression.

(361)   428.5 + 379.5

(362)   399.2 + 108.4

(363)   225.6 − 141.3

(364)   434.2 + 86.5

(365)   288.53 + 426.817

(366)   313.7 − 15.2

(367)   401 + 379.2

(368)   292.6 + 368.17

(369)   58.1 + 27.3

(370)   266 + 219.3

DECIMALS

Evaluate each expression.

(371) 300.027 − 106.72

(372) 431.8 + 8.9

(373) 284.9 + 480.6

(374) 318.71 + 242.8

(375) 482.9 − 1.4

(376) 11.4 + 185.6

(377) 396.5 + 82.2

(378) 414.76 − 206.9

(379) 353.4 − 336.4

(380) 446.5 + 495.4

DECIMALS

Evaluate each expression.

(381)   494.17 + 402.2

(382)   365.9 + 206.5

(383)   391.7 + 52.8

(384)   337.5 + 376.7

(385)   494.6 − 53.6

(386)   463.6 − 277.2

(387)   390.09 + 257.4

(388)   177.4 + 150.6

(389)   458.5 + 312.6

(390)   180.2 + 186.8

DECIMALS

Find the product of the below decimals.

(391)   11.6 × 6.3

(392)   9 × 2.7

(393)   3.3 × 10.6

(394)   6.839 × 1.6

(395)   11 × 13.4

(396)   12.2 × 3.7

(397)   10.21 × 7.5

(398)   11.67 × 5.86

(399)   1.9 × 11.2

(400)   2.9 × 13.3

DECIMALS

Find the product of the below decimals.

(401) 1.6 × 6.5

(402) 5.7 × 1.3

(403) 11.7 × 5.5

(404) 4.4 × 0.1

(405) 11.286 × 0.82

(406) 12.3 × 11.38

(407) 12.1 × 2.8

(408) 5.7 × 9.1

(409) 12.2 × 6.7

(410) 9.3 × 6.7

DECIMALS

Basic Math

Find the product of the below decimals.

(411) 3.4 × 0.67

(412) 5 × 3.71

(413) 7.3 × 4.9

(414) 4.4 × 1.1

(415) 12.7 × 10.35

(416) 7.5 × 12.2

(417) 4 × 12.6

(418) 0.96 × 2.64

(419) 6.6 × 13.5

(420) 10.3 × 7.7

DECIMALS

Basic Math

Find the product of the below decimals.

(421) 0.2 × 8.3

(422) 11.2 × 6.2

(423) 7.5 × 8.5

(424) 9.6 × 0.5

(425) 3.9 × 13.56

(426) 3.1 × 2.2

(427) 1.7 × 3.13

(428) 7 × 0.8

(429) 7.8 × 8.7

(430) 5.21 × 7.1

DECIMALS

Find the product of the below decimals.

(431) 1.3 × 8.1

(432) 12.3 × 0.2

(433) 2.8 × 12

(434) 13.93 × 10.19

(435) 10.1 × 7.5

(436) 10.9 × 12.452

(437) 1.3 × 7.7

(438) 0.8 × 7.4

(439) 12 × 7.87

(440) 7.3 × 13.6

DECIMALS

Basic Math

Find the product of the below decimals.

(441) 48.442 × 35.8

(442) 31.3 × 15.7

(443) 29 × 31

(444) 15.1 × 12.7

(445) 22.4 × 1.8

(446) 41.1 × 41

(447) 39.2 × 39.4

(448) 27.9 × 30.6

(449) 27.3 × 41.36

(450) 2.9 × 46.5

DECIMALS

Find the product of the below decimals.

(451) 26.3 × 6.3

(452) 29.8 × 36.9

(453) 38.3 × 7.1

(454) 39 × 12.5

(455) 32.36 × 38.2

(456) 40.7 × 25.5

(457) 3.1 × 0.5

(458) 0.9 × 36.2

(459) 5 × 8.132

(460) 6.2 × 49

DECIMALS

Basic Math

Find the product of the below decimals.

(461) 14.5 × 1.7

(462) 9.865 × 34.97

(463) 21.9 × 21.6

(464) 46.5 × 27.9

(465) 33.4 × 31.3

(466) 7.5 × 24.1

(467) 11 × 33.9

(468) 8.6 × 26.6

(469) 15.49 × 3.2

(470) 44.3 × 33.4

DECIMALS

Find the product of the below decimals.

(471) 38.8 × 33.197

(472) 16.3 × 1.9

(473) 49.93 × 17.5

(474) 14.9 × 22.2

(475) 42.7 × 33.4

(476) 29.6 × 2.5

(477) 4.2 × 38.6

(478) 35.5 × 27.7

(479) 44.97 × 36.37

(480) 2.2 × 25.6

DECIMALS

Find the product of the below decimals.

(481) 47.07 × 4.3

(482) 41.2 × 46.8

(483) 25.3 × 45.8

(484) 0.5 × 29.6

(485) 36.9 × 41.5

(486) 44.7 × 43.5

(487) 35.2 × 47

(488) 34 × 39

(489) 20 × 1.8

(490) 10 × 27.09

DECIMALS

Find the product of the below decimals.

(491)   20 × 17 × 13.4

(492)   7.4 × 2.39 × 11.8

(493)   13 × 6.895 × 11.1

(494)   13 × 3.6 × 19.4

(495)   18.5 × 9.6 × 16.7

(496)   14.8 × 16.56 × 12.68

(497)   1.8 × 8.8 × 5.336

(498)   2.4 × 9.55 × 12.143

(499)   16 × 7.09 × 13.2

(500)   2.9 × 9.2 × 14.3

DECIMALS

Basic Math

Find the product of the below decimals.

(501) 8.5 × 0.9 × 15.4

(502) 4.8 × 14.8 × 14.6

(503) 1.336 × 12.1 × 18.5

(504) 14.169 × 4.8 × 4.6

(505) 12.92 × 6.895 × 12.943

(506) 14.2 × 4.8 × 1.7

(507) 17.1 × 3.3 × 13.8

(508) 2.2 × 1.9 × 18.7

(509) 17.8 × 2.9 × 9.2

(510) 15.7 × 5.6 × 0.45

DECIMALS

Find the product of the below decimals.

(511)   13.8 × 19.1 × 0.8

(512)   18.7 × 2.5 × 0.07

(513)   12.7 × 0.8 × 1.65

(514)   16.9 × 8 × 4.7

(515)   8.9 × 10.4 × 18.3

(516)   12.943 × 2.9 × 1.9

(517)   5.5 × 6.9 × 14.8

(518)   4 × 3.5 × 17.3

(519)   15.8 × 7.6 × 11.98

(520)   10.1 × 1.3 × 6.7

DECIMALS

Basic Math

Find the product of the below decimals.

(521)   11.5 × 15.8 × 5.9

(522)   12.1 × 12.16 × 5.4

(523)   12.2 × 3.5 × 14.9

(524)   4.5 × 7.2 × 2.5

(525)   15 × 14.3 × 6.3

(526)   17.7 × 7.2 × 19.6

(527)   0.5 × 7.8 × 19.14

(528)   19.1 × 3.2 × 4.2

(529)   17.41 × 1.9 × 17.7

(530)   7.1 × 15.8 × 0.7

DECIMALS

Find the product of the below decimals.

(531)   7.07 × 1.6 × 13.7

(532)   1.5 × 9.16 × 6.5

(533)   8.3 × 1.5 × 12

(534)   13.1 × 5 × 6.2

(535)   6.5 × 0.7 × 7.8

536)   2.4 × 0.6 × 5.3

(537)   6.3 × 0.7 × 7.8

(538)   19.5 × 10.7 × 13.8

(539)   15.22 × 15.8 × 1.4

(540)   19.6 × 3.2 × 18.1

DECIMALS

Find the quotient of the below decimals.

(541) 4.8 ÷ 8

(542) 11.9 ÷ 3.4

(543) 7.6 ÷ 0.2

(544) 11.5 ÷ 11.5

(545) 7.1 ÷ 5

(546) 8.4 ÷ 3

(547) 4.5 ÷ 0.9

(548) 11.9 ÷ 6.8

(549) 2.6 ÷ 13

(550) 2.9 ÷ 14.5

DECIMALS

Find the quotient of the below decimals.

(551) 10.8 ÷ 12

(552) 14.5 ÷ 0.5

(553) 13 ÷ 2.5

(554) 3.5 ÷ 2

(555) 8.4 ÷ 4.8

(556) 1 ÷ 2.5

(557) 12.2 ÷ 0.4

(558) 6 ÷ 0.8

(559) 4.32 ÷ 3

(560) 9.4 ÷ 4.7

DECIMALS

Basic Math

Find the quotient of the below decimals.

(561) 14.1 ÷ 1.2 (562) 8.4 ÷ 8

(563) 5.2 ÷ 10.4 (564) 14.2 ÷ 0.2

(565) 14.9 ÷ 0.1 (566) 1.68 ÷ 11.2

(567) 3.9 ÷ 1.3 (568) 14.9 ÷ 5

(569) 13.9 ÷ 2.5 (570) 11.7 ÷ 0.2

DECIMALS

Find the quotient of the below decimals.

(571) 10.4 ÷ 2

(572) 9.3 ÷ 0.4

(573) 3.6 ÷ 7.2

(574) 3.3 ÷ 1.1

(575) 7.5 ÷ 5

(576) 12.2 ÷ 0.8

(577) 12 ÷ 3.2

(578) 11.2 ÷ 0.2

(579) 5.8 ÷ 11.6

(580) 8.4 ÷ 11.2

DECIMALS

Find the quotient of the below decimals.

(581) 11.7 ÷ 9.75

(582) 6.4 ÷ 4

(583) 9.3 ÷ 6

(584) 2.7 ÷ 3.6

(585) 3.6 ÷ 1.2

(586) 1.5 ÷ 2.4

(587) 8.4 ÷ 4

(588) 11.7 ÷ 0.4

(589) 3 ÷ 2

(590) 7.581 ÷ 5.7

DECIMALS

Find the quotient of the below decimals.

(591) 92.4 ÷ 4.2

(592) 61.8 ÷ 30.9

(593) 63.6 ÷ 12

(594) 38.1 ÷ 6

(595) 24 ÷ 3.2

(596) 1.3 ÷ 0.1

(597) 94 ÷ 4

(598) 9.4 ÷ 2

(599) 12 ÷ 0.8

(600) 17.4 ÷ 12

DECIMALS

Find the quotient of the below decimals.

(601) 20.8 ÷ 0.8

(602) 22.6 ÷ 10

(603) 88.2 ÷ 4.2

(604) 35.7 ÷ 42

(605) 58 ÷ 14.5

(606) 2.7 ÷ 0.6

(607) 47 ÷ 2

(608) 78 ÷ 20

(609) 57 ÷ 0.6

(610) 43.2 ÷ 22.5

DECIMALS

Find the quotient of the below decimals.

(611) 10.5 ÷ 2

(612) 37.4 ÷ 0.8

(613) 65.4 ÷ 0.8

(614) 18.3 ÷ 0.3

(615) 32.2 ÷ 35

(616) 12 ÷ 10

(617) 43.5 ÷ 1.2

(618) 81.9 ÷ 0.4

(619) 51.2 ÷ 80

(620) 37 ÷ 12.5

DECIMALS

Find the quotient of the below decimals.

(621) 78 ÷ 0.39

(622) 88.4 ÷ 0.5

(623) 34.5 ÷ 50

(624) 87.6 ÷ 10

(625) 9.8 ÷ 3.5

(626) 72.8 ÷ 56

(627) 75.3 ÷ 0.6

(628) 48.4 ÷ 88

(629) 17.5 ÷ 2.8

(630) 13.7 ÷ 25

DECIMALS

Find the quotient of the below decimals.

(631) 70.6 ÷ 4

(632) 83.7 ÷ 10

(633) 90.5 ÷ 12.5

(634) 67.5 ÷ 0.5

(635) 50.8 ÷ 20

(636) 57.72 ÷ 4

(637) 68.9 ÷ 2.6

(638) 34.5 ÷ 75

(639) 12.6 ÷ 18

(640) 30.3 ÷ 0.6

DECIMALS

Basic Math

Round each of the below decimals to the place underlined.

(641) 1<u>4</u>6

(642) <u>6</u>.9

(643) 22,2<u>6</u>8

(644) 8.<u>9</u>8

(645) 6.<u>2</u>48

(646) 5,00<u>9</u>.49

(647) 1.<u>1</u>5

(648) 99<u>2</u>.771

(649) 78.<u>9</u>726

(650) 1.<u>9</u>88

DECIMALS

Round each of the below decimals to the place underlined.

(651)   25<u>0</u>.2

(652)   72.<u>9</u>35

(653)   86<u>6</u>.35

(654)   8.<u>8</u>28

(655)   <u>9</u>.8

(656)   29,80<u>7</u>.3

(657)   40.<u>9</u>78

(658)   521,3<u>7</u>4.272

(659)   60<u>8</u>.918

(660)   3.<u>5</u>221

DECIMALS

Round each of the below decimals to the place underlined.

(661) 476,789.188   (662) 7.320

(663) 212,486   (664) 4.071

(665) 86.9704   (666) 7.7953

(667) 4.2977   (668) 3.11538

(669) 3.914   (670) 8.969

DECIMALS

Basic Math

Round each of the below decimals to the place underlined.

(671)   0.1<u>8</u>00

(672)   0.<u>1</u>149

(673)   9.<u>5</u>08

(674)   2.<u>8</u>340

(675)   8.<u>6</u>995

(676)   3.0<u>9</u>3

(677)   4.9<u>9</u>4

(678)   7.<u>0</u>20

(679)   0.<u>1</u>492

(680)   6.<u>7</u>71

DECIMALS

Basic Math

Round each of the below decimals to the place underlined.

(681) 2.97

(682) 1.9460

(683) 2.84817

(684) 2.9947

(685) 7.2141

(686) 3.88970

(687) 6.804

(688) 9.55

(689) 2.9787

(690) 5.1555

DECIMALS

Round each of the below decimals to the place underlined.

(691) 5.521<u>9</u>39

(692) 9.059<u>9</u>4

(693) 6.410<u>1</u>00

(694) 4.43<u>8</u>4

(695) 4.196<u>0</u>4

(696) 9.23<u>0</u>71

(697) 2.843<u>3</u>62

(698) 5.467<u>3</u>4

(699) 6.461<u>7</u>40

(700) 7.167<u>5</u>92

DECIMALS

Basic Math

Round each of the below decimals to the place underlined.

(701) 1.14<u>9</u>5

(702) 4.70<u>2</u>48

(703) 5.06<u>9</u>0

(704) 9.368<u>3</u>1

(705) 9.651<u>1</u>1

(706) 7.95<u>2</u>37

(707) 1.887<u>9</u>5

(708) 3.879<u>7</u>3

(709) 6.28<u>9</u>877

(710) 2.80<u>6</u>043

DECIMALS

Round each of the below decimals to the place underlined.

(711) 8.491<u>9</u>96

(712) 9.73<u>4</u>325

(713) 5.24<u>4</u>1

(714) 3.342<u>7</u>51

(715) 4.897<u>4</u>18

(716) 4.6492<u>6</u>85

(717) 9.09014<u>4</u>2

(718) 8.54661<u>9</u>8

(719) 1.91028<u>4</u>2

(720) 8.5612<u>0</u>6

DECIMALS

Basic Math

Round each of the below decimals to the place underlined.

(721)   9.7812<u>5</u>6

(722)   8.24277<u>4</u>4

(723)   3.62454<u>1</u>1

(724)   1.2444<u>3</u>1

(725)   2.9556<u>9</u>4

(726)   7.9853<u>9</u>0

(727)   4.1945<u>1</u>91

(728)   5.4225<u>2</u>0

(729)   9.11185<u>9</u>9

(730)   4.58780<u>0</u>0

DECIMALS

Round each of the below decimals to the place underlined.

(731) 5.1574<u>9</u>12

(732) 2.1804<u>9</u>40

(733) 6.5556<u>6</u>3

(734) 4.20861<u>9</u>4

(735) 9.1086<u>9</u>0

(736) 1.09502<u>9</u>7

(737) 9.90081<u>5</u>3

(738) 8.5655<u>4</u>60

(739) 3.4366<u>9</u>2

(740) 1.2595<u>0</u>4

DECIMALS

Write each of the below numeral in words.

(741) 0.4

(742) 0.9

(743) 0.2

(744) 0.3

(745) 0.5

(746) 0.49

(747) 0.52

(748) 0.07

(749) 0.06

(750) 0.55

DECIMALS

Write each of the below numeral in words.

(751) 0.08

(752) 0.96

(753) 0.32

(754) 0.03

(755) 0.75

(756) 0.621

(757) 0.216

(758) 0.301

(759) 0.555

(760) 0.027

DECIMALS

Basic Math

Write each of the below numeral in words.

(761) 0.008

(762) 0.756

(763) 0.086

(764) 0.001

(765) 0.597

(766) 0.0908

(767) 0.0574

(768) 0.4742

(769) 0.0848

(770) 0.0402

DECIMALS

Basic Math

Write each of the below numeral in words.

(771) 0.0001

(772) 0.0707

(773) 0.0064

(774) 0.4002

(775) 0.4022

(776) 0.09793

(777) 0.06975

(778) 0.01282

(779) 0.04005

(780) 0.03509

DECIMALS

Write each of the below numeral in words.

(781) 0.15338

(782) 0.87078

(783) 0.20819

(784) 0.00077

(785) 0.27304

(786) 0.309503

(787) 0.025008

(788) 0.730304

(789) 0.000493

(790) 0.219003

DECIMALS

Write each of the below numeral in words.

(791) 0.000001

(792) 0.000508

(793) 0.000107

(794) 0.900029

(795) 0.024908

(796) 0.0000006

(797) 0.0060048

(798) 0.7020007

(799) 0.8690007

(800) 0.0043457

DECIMALS

Basic Math

Write each of the below numeral in words.

(801) 0.0000031

(802) 0.0000008

(803) 0.0570289

(804) 0.0904005

(805) 0.4518773

(806) 0.30360224

(807) 0.09315604

(808) 0.26000808

(809) 0.70090106

(810) 0.60008038

DECIMALS

Write each of the below numeral in words.

(811)   0.62300003

(812)   0.50000048

(813)   0.70006025

(814)   0.43190009

(815)   0.10022313

(816)   0.603008064

(817)   0.095232081

(818)   0.163003001

(819)   0.676110094

(820)   0.800150201

Write each of the below numeral in words.

(821) 0.486607106

(822) 0.200103008

(823) 0.106008803

(824) 0.500806086

(825) 0.361605063

DECIMALS

Write each of the below words as a numeral.

(826) two tenths

(827) four tenths

(828) three tenths

(829) eight tenths

(830) seven tenths

(831) two hundred ten and sixty-four hundredths

(832) eight hundred sixty and twenty-seven hundredths

(833) five hundred and four hundredths

(834) five hundred fifty and ninety-eight hundredths

(835) seven hundred and sixty-two hundredths

DECIMALS

Write each of the below words as a numeral.

(836)   one hundred fifty and seventy-three hundredths

(837)   one hundred ninety-seven and six hundredths

(838)   one hundred sixty-one and ninety-eight hundredths

(839)   three hundred and two hundredths

(840)   one hundred seven and nine hundredths

DECIMALS

Write each of the below words as a numeral.

(841)   one hundred five and one hundredth

(842)   one hundred thirty and thirty-three hundredths

(843)   three hundred five and thirty-five hundredths

(844)   seven hundred six and eighty-eight hundredths

(845)   six hundred ten and seventy-six hundredths

DECIMALS

Write each of the below words as a numeral.

(846) three hundred nine and forty-eight hundredths

(847) four hundred six and eighty-one hundredths

(848) five hundred ninety and five hundredths

(849) seven hundred and three hundredths

(850) six hundred ninety-one and eight hundredths

DECIMALS

Write each of the below words as a numeral.

(851) six hundred seven and nine thousandths

(852) nine hundred eight and eighty-four thousandths

(853) one hundred fifty-four and seven hundred eight thousandths

(854) nine hundred seventy and ninety-two thousandths

(855) five hundred fifty and fifty-eight thousandths

DECIMALS

Write each of the below words as a numeral.

(856)　two hundred seventy and seventy-one thousandths

(857)　four hundred seven and eight hundred seventy-six thousandths

(858)　eight hundred two and sixty-seven thousandths

(859)　six hundred forty-eight and six hundred six thousandths

(860)　two hundred fifty-four and four hundred eight thousandths

DECIMALS

Write each of the below words as a numeral.

(861)   six hundred and five hundred ninety-nine thousandths

(862)   six hundred fifty-seven and ninety-six thousandths

(863)   seven hundred nine and five hundred sixty-eight thousandths

(864)   six hundred and three hundred forty-nine thousandths

(865)   three hundred thirty and five hundred eight thousandths

DECIMALS

Write each of the below words as a numeral.

(866)   seven hundred fifty-four and two thousandths

(867)   nine hundred sixty-seven and five hundred six thousandths

(868)   four hundred ninety and twenty-seven thousandths

(869)   nine hundred sixty-nine and fifty-one thousandths

(870)   eight hundred and eight hundred five thousandths

DECIMALS

Write each of the below words as a numeral.

(871) seventy-nine million, three hundred eighty thousand, forty-six and nine thousand, one hundred fifty-three ten-thousandths

(872) nine hundred sixty-four million, seven hundred eight thousand, nine hundred ninety and four thousand, five hundred fourteen ten-thousandths

(873) eight hundred forty-five million, five hundred forty thousand, three hundred five and one thousand, four ten-thousandths

(874) four million, forty thousand, one hundred three and nine ten-thousandths

(875) two million, eight hundred forty-three thousand, one hundred eighty-seven and six thousand, one hundred fifty-five ten-thousandths

DECIMALS

Write each of the below words as a numeral.

(876) two million, eight hundred fifty-six thousand, six hundred fifty and one thousand, three hundred three ten-thousandths

(877) four million, six hundred thousand, eighty-three and five thousand, three hundred eighty-three ten-thousandths

(878) five hundred eighty million, sixty-eight thousand, four hundred five and four thousand, four hundred eighty-two ten-thousandths

(879) eighty-nine million, one hundred one thousand, three hundred six and four thousand, seven ten-thousandths

(880) eighty-nine million, three hundred forty thousand, eight hundred thirty-three and eight thousand, ninety-eight ten-thousandths

DECIMALS

Write each of the below words as a numeral.

(881) one hundred three million, six thousand, eight hundred and eight thousand, nine hundred five ten-thousandths

(882) five million, four hundred six thousand, three hundred and three thousand, four ten-thousandths

(883) nine million, twenty and nine hundred seventy-five ten-thousandths

(884) twenty-three million, seven hundred one thousand, twenty and one thousand, one hundred fifty-nine ten-thousandths

(885) seventy million, eighty-one thousand, nine and six thousand, thirteen ten-thousandths

DECIMALS

Write each of the below words as a numeral.

(886) one hundred four million, three hundred three thousand, one hundred sixty and five thousand, thirty-three ten-thousandths

(887) eight million, six hundred nine thousand, eighty and nine hundred seven ten-thousandths

(888) four hundred twenty million, nine thousand, thirty and eight thousand, nine hundred thirteen ten-thousandths

(889) three million, seven hundred thousand, ninety and seven hundred thirty-nine ten-thousandths

(890) eight million, five hundred thousand, thirteen and seven thousand, six ten-thousandths

DECIMALS

Write each of the below words as a numeral.

(891) sixteen million, nine hundred fifty-nine thousand, two hundred and nine thousand, four hundred twenty-five hundred-thousandths

(892) one hundred million, one hundred eighty-six thousand, eight hundred seventy and two hundred-thousandths

(893) one hundred forty-four million, one hundred twenty-four thousand, six hundred fifty and six thousand, ninety-two hundred-thousandths

(894) sixty million, four hundred ninety-five thousand, five hundred and eighty thousand, ninety-one hundred-thousandths

(895) five hundred three million, nine hundred fifty and three thousand, eight hundred sixty-six hundred-thousandths

DECIMALS

Write each of the below words as a numeral.

(896) ninety-nine million, seven hundred fifty thousand, six hundred and one thousand, two hundred seventeen hundred-thousandths

(897) six hundred five million, eight hundred seventy-seven thousand, two and seventy-nine thousand, seven hundred eight hundred-thousandths

(898) six hundred eight million, one hundred seven thousand, six and forty-one thousand, one hundred six hundred-thousandths

(899) ten million, twenty-five and four hundred seventy-nine hundred-thousandths

(900) forty-two million, fifty thousand, eighty-eight and six thousand, nine hundred one hundred-thousandths

DECIMALS

Write each of the below words as a numeral.

(901) four million, four hundred thousand, one hundred and four thousand, three hundred nine hundred-thousandths

(902) twenty million, three hundred two thousand, two and six thousand, eight hundred forty-two hundred-thousandths

(903) nine hundred million, two hundred seventy thousand, eight hundred six and forty-eight thousand, one hundred nine hundred-thousandths

(904) twenty-six million, six hundred forty-nine thousand, two hundred and ninety-one thousand, thirty-one hundred-thousandths

(905) fifty-eight million, four hundred twenty thousand, fifty and seventy thousand, seven hundred twenty-three hundred-thousandths

Write each of the below words as a numeral.

(906) two hundred million, one hundred eight thousand, six hundred twenty and five thousand, eight hundred-thousandths

(907) five million, seven hundred five thousand, nine hundred twenty-four and five hundred three hundred-thousandths

(908) ninety-four million, six hundred sixty-six thousand, one hundred and thirty-eight thousand, two hundred-thousandths

(909) nine hundred twenty-eight million, three hundred thirty-one thousand and eighty-two hundred-thousandths

(910) six hundred seventy-four million, five thousand, four and forty thousand, three hundred five hundred-thousandths

DECIMALS

Write each of the below words as a numeral.

(911) five hundred eighty thousand, one hundred eighty and three million, eighty thousand, six hundred sixty-three ten-millionths

(912) eight thousand, forty-three and three million, seven hundred one ten-millionths

(913) eight hundred ninety-seven thousand and four million, sixty-three thousand, seven ten-millionths

(914) three thousand, thirteen and seven hundred twenty-five thousand, seventy-six ten-millionths

(915) nine hundred eighty thousand, three hundred twenty and one million, two hundred twenty thousand, ninety-five ten-millionths

DECIMALS

Write each of the below words as a numeral.

(916) seven thousand, forty and seven million, fifty-seven thousand, eight ten-millionths

(917) three hundred thousand, twenty and seven million, five hundred thousand, two hundred nine ten-millionths

(918) ninety thousand, eighty-four and ninety-eight thousand, forty-nine ten-millionths

(919) twenty thousand, six hundred eighty-one and seven hundred eight thousand, three ten-millionths

(920) twenty thousand and three million, six hundred twenty-seven ten-millionths

DECIMALS

Write each of the below words as a numeral.

(921) seven hundred fifty-one thousand, eight hundred forty and eight million, one hundred eighty-six thousand, two hundred two ten-millionths

(922) sixty-nine thousand, three hundred and six million, two hundred four thousand, four hundred fifty-three ten-millionths

(923) seventy-six thousand, four and eight hundred seventy thousand, seventeen ten-millionths

(924) six hundred sixty-two thousand, seventy-seven and six million, four hundred thousand, seventy-two ten-millionths

(925) seven thousand, six hundred and two hundred thousand, four ten-millionths

DECIMALS

Write each of the below words as a numeral.

(926) eight thousand, six and eight million, nine hundred fifty-one ten-millionths

(927) eighteen thousand, forty and six million, three hundred seven ten-millionths

(928) nine thousand, four and nine hundred twenty-one thousand, nine hundred three ten-millionths

(929) six hundred eight thousand, nine hundred twenty-four and thirty thousand, eighty-three ten-millionths

(930) three thousand, ninety and two million, eight hundred ninety-one thousand, four hundred eighty-seven ten-millionths

DECIMALS

Write each of the below words as a numeral.

(931) seven billion, two hundred ninety million, fifty-three thousand, one hundred fifty and four million, sixty-three thousand, seven hundred one hundred-millionths

(932) sixty-five billion, two hundred eight million, three hundred four thousand, nine hundred eighty and thirty-seven million, seven thousand, five hundred nine hundred-millionths

(933) three billion, eighty-five million, eighty thousand, seven hundred ninety-one and twenty million, nine hundred sixty thousand, ninety-five hundred-millionths

(934) seventy billion, four million, two hundred four thousand, six hundred three and sixty million, forty-three thousand, twenty-four hundred-millionths

(935) eighty-nine billion, seventy-five million, eight thousand, ninety-six and four million, six hundred thousand, six hundred-millionths

DECIMALS

Basic Math

Write each of the below words as a numeral.

(936) one hundred forty-one billion, three hundred ten and three hundred forty thousand, forty-seven hundred-millionths

(937) four billion, three hundred twenty million, one thousand, eighty-five and eight million, seventy-nine thousand, one hundred seventy-eight hundred-millionths

(938) two hundred thirty billion, one million, three hundred nine thousand, six hundred fifty-three and seventy million, five hundred eight thousand, seven hundred five hundred-millionths

(939) sixty billion, three hundred twenty million, two hundred nine thousand, one and two hundred one thousand, five hundred-millionths

(940) seventy-five billion, nine hundred four million, five hundred seventy thousand, five hundred forty and thirteen million, twenty-two thousand, seven hundred-millionths

DECIMALS

Write each of the below words as a numeral.

(941) nine billion, three hundred ten million, eight hundred six and ninety-nine million, seventy-nine thousand, four hundred-millionths

(942) seven billion, one hundred nine million, seven hundred thousand, ten and sixty-one million, one hundred thousand, ninety-five hundred-millionths

(943) three hundred eighty-one billion, two hundred seventy-one million, seven hundred forty thousand, one hundred and seven hundred thousand, six hundred-millionths

(944) forty billion, thirty million, nine hundred ninety-eight thousand, two hundred eighty-three and seventy-seven million, nine thousand, forty-one hundred-millionths

(945) five billion, ninety-seven million, three thousand, seven hundred sixty and eight million, eight hundred fifty thousand, eight hundred-millionths

DECIMALS

Write each of the below words as a numeral.

(946) ninety billion, seven hundred twenty million, eight hundred thirty thousand, eighty and ten million, six hundred forty-six hundred-millionths

(947) fifty-six billion, one million, eighty-five thousand, eighty and seventy-nine million, one hundred seventy-nine thousand, six hundred seven hundred-millionths

(948) ninety-one billion, six hundred million, seventy thousand, five hundred and forty-four million, eight thousand, one hundred nine hundred-millionths

(949) one hundred eighty billion, ten million, four hundred sixty thousand, six hundred ninety-nine and eighty million, three hundred thousand, four hundred three hundred-millionths

(950) seventy billion, nine hundred three million, ninety thousand, four hundred ninety and six million, eight hundred thirty thousand, three hundred eighty-two hundred-millionths

DECIMALS

Write the name of the decimal place as underlined for each of the below decimals.

(951) 2.<u>7</u>7

(952) 48.0<u>9</u>8

(953) 75.<u>7</u>5

(954) 4.<u>5</u>350

(955) 42<u>2</u>

(956) 39,90<u>9</u>.182

(957) <u>1</u>.9

(958) 1.<u>8</u>719

(959) 76.<u>4</u>22

(960) 61.2<u>7</u>2

DECIMALS

Basic Math

Write the name of the decimal place as underlined for each of the below decimals.

(961)   8,48<u>1</u>.0

(962)   10,77<u>1</u>

(963)   8.9<u>0</u>0

(964)   72.<u>3</u>02

(965)   2.<u>7</u>804

(966)   8.<u>3</u>0

(967)   9.0<u>7</u>4

(968)   80,68<u>7</u>.194

(969)   24<u>1</u>.2

(970)   96.<u>5</u>71

DECIMALS

Write the name of the decimal place as underlined for each of the below decimals.

(971) 8.<u>8</u>022

(972) 83<u>4</u>.9

(973) 22,51<u>3</u>

(974) 1.4<u>5</u>22

(975) 5.5<u>0</u>73

(976) 13.7<u>2</u>36

(977) 83,18<u>8</u>

(978) 0.6<u>7</u>

(979) 844,75<u>2</u>.989

(980) 9.<u>8</u>60

DECIMALS

Basic Math

Write the name of the decimal place as underlined for each of the below decimals.

(981) 3,92<u>3</u>.086

(982) 300,69<u>8</u>

(983) 58<u>8</u>

(984) 1<u>2</u>

(985) <u>6</u>

(986) 48.1<u>6</u>5

(987) 9.8<u>7</u>3

(988) 3<u>2</u>

(989) 22<u>4</u>.04

(990) 6.9<u>9</u>1

DECIMALS

Write the name of the decimal place as underlined for each of the below decimals.

(991)   25.4<u>3</u>61

(992)   8.<u>3</u>92

(993)   4.7<u>8</u>615

(994)   2.3<u>5</u>532

(995)   2.<u>4</u>7

(996)   8.<u>3</u>664

(997)   84.8<u>3</u>47

(998)   3<u>7</u>

(999)   43.<u>7</u>9

(1000)   <u>3</u>.92

DECIMALS

Basic Math

Write the name of the decimal place as underlined for each of the below decimals.

1001) 2.8<u>3</u>24

(1002) 8.16<u>9</u>5

(1003) 2.9<u>5</u>511

(1004) 5.72<u>5</u>94

(1005) 2.2<u>8</u>314

(1006) 8.4<u>3</u>7

(1007) 4.67<u>5</u>4

(1008) 4.0<u>8</u>3

(1009) 3.37<u>9</u>90

(1010) 3.21<u>0</u>02

DECIMALS

Write the name of the decimal place as underlined for each of the below decimals.

(1011)   6.01<u>8</u>398

(1012)   6.1<u>2</u>050

(1013)   0.03<u>6</u>98

(1014)   0.4<u>7</u>07

(1015)   4.9<u>1</u>349

(1016)   0.53<u>7</u>28

(1017)   9.14<u>7</u>697

(1018)   6.75<u>2</u>25

(1019)   0.08<u>8</u>106

(1020)   6.5<u>7</u>892

DECIMALS

Basic Math

Write the name of the decimal place as underlined for each of the below decimals.

(1021) 4.0<u>7</u>2

(1022) 6.8<u>4</u>91

(1023) 0.90<u>2</u>3

(1024) 7.5<u>2</u>854

(1025) 4.64<u>2</u>37

(1026) 5.2<u>7</u>0

(1027) 4.94<u>5</u>49

(1028) 2.4<u>3</u>01

(1029) 7.21<u>4</u>260

(1030) 5.2<u>4</u>8

DECIMALS

Basic Math

Write the name of the decimal place as underlined for each of the below decimals.

(1031) 5.1<u>1</u>294   (1032) 9.72<u>1</u>31

(1033) 3.0<u>1</u>8   (1034) 9.69<u>4</u>647

(1035) 5.8<u>7</u>093   (1036) 6.9<u>6</u>86

(1037) 3.8<u>7</u>8   (1038) 9.79<u>6</u>5

(1039) 5.3<u>0</u>948   (1040) 3.1<u>0</u>35

DECIMALS

Basic Math

Write the name of the decimal place as underlined for each of the below decimals.

(1041) 5.50<u>1</u>8

(1042) 6.1<u>8</u>391

(1043) 1.231<u>5</u>

(1044) 2.3<u>6</u>66

(1045) 0.16<u>6</u>8

(1046) 3.11<u>6</u>8

(1047) 8.9<u>5</u>640

(1048) 6.8<u>1</u>9

(1049) 6.3<u>1</u>774

(1050) 2.17<u>9</u>793

DECIMALS

Basic Math

Write the name of the decimal place as underlined for each of the below decimals.

(1051) 2.6626<u>2</u>56   (1052) 9.402<u>2</u>64

(1053) 1.452<u>8</u>07   (1054) 0.790<u>3</u>96

(1055) 2.413<u>5</u>6   (1056) 0.618<u>0</u>79

(1057) 3.651<u>5</u>5   (1058) 6.658<u>0</u>50

(1059) 9.96718<u>6</u>4   (1060) 2.761<u>3</u>42

DECIMALS

Basic Math

Write the name of the decimal place as underlined for each of the below decimals.

(1061) 3.5857<u>5</u>99

(1062) 2.005<u>8</u> 8

(1063) 3.927<u>1</u>4

(1064) 3.39442 <u>8</u>2

(1065) 0.809<u>4</u>973

(1066) 0.00067 <u>2</u>9

(1067) 5.780<u>7</u>9

(1068) 4.308<u>4</u> 41

(1069) 3.72912<u>7</u>2

(1070) 2.327<u>8</u> 8

DECIMALS

Basic Math

Write the name of the decimal place as underlined for each of the below decimals.

(1071)   8.02981 4

(1072)   3.498764

(1073)   5.05000 27

(1074)   5.515490

(1075)   7.616672 8

(1076)   2.836420

(1077)   1.83872 4 9

(1078)   0.616641

(1079)   7.221272

(1080)   6.7926784

DECIMALS

Basic Math

Write the name of the decimal place as underlined for each of the below decimals.

(1081) 3.101<u>0</u>47

(1082) 6.144<u>1</u>4

(1083) 5.4670<u>1</u>8

(1084) 9.93643<u>6</u>8

(1085) 2.286<u>4</u>86

(1086) 0.108<u>8</u>8

(1087) 0.44771<u>1</u>9

(1088) 3.9423<u>3</u>96

(1089) 0.594<u>6</u>2

(1090) 2.363<u>6</u>44

DECIMALS

Write the name of the decimal place as underlined for each of the below decimals.

(1091)   4.57858 <u>2</u>7   (1092)   0.318 <u>2</u>7

(1093)   5.60928 <u>5</u>1   (1094)   8.313 <u>6</u>8

(1095)   5.296 <u>3</u>73   (1096)   5.9461 0<u>4</u>

(1097)   9.473<u>9</u>7   (1098)   8.49871 <u>4</u>4

(1099)   4.1892 <u>9</u>2   (1100)   4.8909 <u>3</u>9

DECIMALS

Round each of the below decimals to the place indicated.

(1101) 17.25 ; tenths

(1102) 15.25 ; ones

(1103) 699.56 ; ones

(1104) 53.681 ; tenths

(1105) 0.9991 ; tenths

(1106) 3.964 ; ones

(1107) 9.297 ; hundredths

(1108) 7.67785 ; hundredths

(1109) 2.4603 ; hundredths

(1110) 30.38739 ; hundredths

DECIMALS

Round each of the below decimals to the place indicated.

(1111) 1.32 ; ones

(1112) 0.9414 ; tenths

(1113) 90.960 ; tenths

(1114) 5.0572 ; tenths

(1115) 9.1 ; ones

(1116) 71,999.1 ; ones

(1117) 46.94292 ; hundredths

(1118) 392.56 ; ones

(1119) 5.7398 ; tenths

(1120) 46.59701 ; hundredths

DECIMALS

Round each of the below decimals to the place indicated.

(1121) 1,329.210 ; ones   (1122) 9.0 ; ones

(1123) 4.8020 ; hundredths   (1124) 5.0814 ; hundredths

(1125) 9.22 ; ones   (1126) 217,594.53 ; ones

(1127) 26.32 ; tenths   (1128) 5.963 ; tenths

(1129) 400,124.61 ; ones   (1130) 37.380 ; tenths

DECIMALS

Basic Math

Round each of the below decimals to the place indicated.

(1131)   5.9259 ; tenths

(1132)   2.8 ; ones

(1133)   1.632 ; tenths

(1134)   7.648 ; hundredths

(1135)   670.81 ; ones

(1136)   69.9890 ; tenths

(1137)   39.4 ; ones

(1138)   69,389.2 ; ones

(1139)   5.39536 ; hundredths

(1140)   6.7978 hundredths

DECIMALS

Basic Math

Round each of the below decimals to the place indicated.

(1141) 7.129 ; tenths

(1142) 439.7 ; ones

(1143) 5.651 ; tenths

(1144) 7.9083 ; tenths

(1145) 8.0207 ; hundredths

(1146) 1.68599 ; hundredths

(1147) 5.94 ; tenths

(1148) 1.566 ; hundredths

(1149) 9.4377 ; tenths

(1150) 9.2864 ; tenths

DECIMALS

Round each of the below decimals to the place indicated.

(1151)   0.3071315 ; hundred-thousandths   (1152)   5.0660153 ; millionths

(1153)   5.655173 ; ten-thousandths   (1154)   8.699687 ; hundred-thousandths

(1155)   3.9532196 ; millionths   (1156)   2.407709 ; thousandths

(1157)   4.3381092 ; millionths   (1158)   4.70019 ; thousandths

(1159)   9.489595 ; thousandths   (1160)   0.65992 ; ten-thousandths

DECIMALS

Round each of the below decimals to the place indicated.

(1161)   1.03931   ; thousandths

(1162)   2.9206085   ; millionths

(1163)   5.56972   ; ten-thousandths

(1164)   2.388809   ; ten-thousandths

(1165)   4.577374   ; hundred-thousandths

(1166)   5.3276694   ; millionths

(1167)   4.049987   ; ten-thousandths

(1168)   8.1002293   ; hundred-thousandths

(1169)   8.73039   ; thousandths

(1170)   1.46907   ; thousandths

DECIMALS

Basic Math

Round each of the below decimals to the place indicated.

(1171)   9.237936 ; hundred-thousandths   (1172)   6.650705 ; thousandths

(1173)   9.261937 ; thousandths   (1174)   3.1356340 ; millionths

(1175)   3.074292 ; thousandths   (1176)   4.9995819 ; hundred-thousandths

(1177)   3.669587 ; thousandths   (1178)   6.778356 ; ten-thousandths

(1179)   1.5871195 ; millionths   (1180)   7.629816 ; ten-thousandths

DECIMALS

Round each of the below decimals to the place indicated.

(1181) 2.1928872 ; hundred-thousandths   (1182) 7.91498 ; ten-thousandths

(1183) 3.566462 ; ten-thousandths   (1184) 6.77544 ; ten-thousandths

(1185) 6.9512050 ; millionths   (1186) 5.2761666 ; millionths

(1187) 6.35919 ; ten-thousandths   (1188) 1.8072172 ; millionths

(1189) 1.86950 ; thousandths   (1190) 1.8100279 ; hundred-thousandths

DECIMALS

Round each of the below decimals to the place indicated.

(1191) 3.166368   ; ten-thousandths   (1192) 3.0125304   ; millionths

(1193) 4.0018   ; thousandths   (1194) 9.6415089   ; millionths

(1195) 6.65653   ; thousandths   (1196) 1.418550   ; hundred-thousandths

(1197) 3.1955195 ; millionths   (1198) 5.218991   ; ten-thousandths

(1199) 5.981649   ; hundred-thousandths (1200) 9.141297   ; ten-thousandths

DECIMALS

Basic Math Answer Keys

DECIMALS

Basic Math Answer Keys

Answer Key

(1) 119.847   (2) 108   (3) 117.3   (4) 120.98

(5) 85.8   (6) 183.2   (7) 103.3   (8) 108.4

(9) 103   (10) 54.7   (11) 82.7   (12) 37.12

(13) 101.35   (14) 102.87   (15) 143.9   (16) 158.6

(17) 14.7   (18) 79.1   (19) 85.6   (20) 160.3

(21) 95.3   (22) 163.7   (23) 45.87   (24) 110.1

(25) 171.17   (26) 97   (27) 184.6   (28) 11.16

(29) 76.5   (30) 74.77   (31) 95.5   (32) 134.2

(33) 59.1   (34) 101.682   (35) 139.6   (36) 144.7

(37) 84.129   (38) 100   (39) 101.6   (40) 162.2

(41) 96.5   (42) 124.3   (43) 44.9   (44) 105.219

(45) 88.08   (46) 116.28   (47) 80.314   (48) 159.64

DECIMALS

Basic Math Answer Keys

(49) 74   (50) 68.5   (51) 716.3   (52) 560

(53) 568.2 (54) 1269.973 (55) 860.9   (56) 746.6

(57) 1186.1 (58) 1283.982 (59) 1313.6   (60) 324.7

(61) 842.9 (62) 1325   (63) 1039.21   (64) 514.9

(65) 971.7 (66) 1392.7   (67) 1405.93   (68) 398.8

(69) 1018.7 (70) 1651.1   (71) 1092.2   (72) 885.178

(73) 758.5 (74) 511   (75) 902   (76) 1295.81

(77) 326.6 (78) 1035.2   (79) 644   (80) 464.4

(81) 1485.9 (82) 1461.5   (83) 686.9   (84) 554.35

(85) 251.4 (86) 642.7   (87) 1354.9   (88) 645.6

(89) 1436.2 (90) 678.2   (91) 1847.947   (92) 1624.3

(93) 1067.609 (94) 421.76 (95) 557.1   (96) 916.9

DECIMALS

Basic Math Answer Keys

(97) 1057.56   (98) 1213.1   (99) 1342   (100) 630.2

(101) 1501.11   (102) 1258.6   (103) 1797.13   (104) 1179.6

(105) 2625.4   (106) 903.5   (107) 2373.8   (108) 1284.3

(109) 1349   (110) 2153.41   (111) 1036.79   (112) 1072.77

(113) 1541   (114) 1346.5   (115) 1596   (116) 865.925

(117) 1188   (118) 1704.21   (119) 812.4   (120) 1409.36

(121) 827.5   (122) 1287.698   (123) 1581.5   (124) 1712.6

(125) 1877   (126) 1460.84   (127) 1450.8   (128) 2168.22

(129) 1611.93   (130) 985.51   (131) 1670.278   (132) 611.36

(133) 1767.8   (134) 1807.56   (135) 772.2   (136) 678.078

(137) 1813.242   (138) 1402.1   (139) 1811.2   (140) 1661.328

(141) 1987.6   (142) 2703.16   (143) 2814.86   (144) 885.8

DECIMALS

Basic Math Answer Keys

(145) 1849.34   (146) 1134.429   (147) 1360.6   (148) 571.79

(149) 1779.3   (150) 1743   (151) 20.55   (152) 28.886

(153) 39.018   (154) 10.65   (155) 3.17   (156) 3.2

(157) 71.11   (158) 81.4   (159) 5.5   (160) 55.3

(161) 11.559   (162) 65.83   (163) 28.9   (164) 7.73

(165) 14.3   (166) 17.6   (167) 9.7   (168) 78.2

(169) 76.2   (170) 8.8   (171) 13.846   (172) 10.6

(173) 56.34   (174) 10.76   (175) 22.1   (176) 21.3

(177) 36.45   (178) 34.3   (179) 77.1   (180) 23.3

(181) 71.5   (182) 43.6   (183) 8.4   (184) 32.7

(185) 5.24   (186) 25.6   (187) 30.04   (188) 19.5

(189) 9.667   (190) 52.666   (191) 12.8   (192) 34.88

DECIMALS

Basic Math Answer Keys

(193) 13.11   (194) 27.9   (195) 34.7   (196) 15.6

(197) 51.71   (198) 32.7   (199) 73.86   (200) 58.7

(201) 35.2   (202) 37.7   (203) 231.8   (204) 31.2

(205) 263.883 (206) 236.6   (207) 155.6   (208) 167.6

(209) 91.9   (210) 80.32   (211) 32   (212) 161.5

(213) 22.2   (214) 195.9   (215) 39   (216) 118.3

(217) 65.16   (218) 104.45 (219) 346.3   (220) 179.872

(221) 339.8   (222) 215.64 (223) 194.346 (224) 398.212

(225) 126.512 (226) 385.9   (227) 144.951 (228) 239.3

(229) 220.74 (230) 135.143 (231) 314.66 (232) 349.42

(233) 22.1   (234) 80.3   (235) 11.83   (236) 203.4

(237) 310.6   (238) 58   (239) 258.32 (240) 214.46

DECIMALS

Basic Math Answer Keys

(241) 18.135   (242) 241.7   (243) 131.34   (244) 307.54

(245) 300.5   (246) 118.2   (247) 147.6   (248) 158.1

(249) 97.1   (250) 141.7   (251) 22.174   (252) 26.857

(253) 39.68   (254) 3.2   (255) 59   (256) 4.5

(257) 16   (258) 63.2   (259) 25.4   (260) 7.53

(261) 21.147   (262) 34.4   (263) 1.99   (264) 10.3

(265) 10.5   (266) 6.29   (267) 82.1   (268) 49.1

(269) 56.6   (270) 3.6   (271) 30.684   (272) 9.2

(273) 16.2   (274) 46.3   (275) 53.3   (276) 9.9

(277) 16.38   (278) 62.088   (279) 15.6   (280) 15.57

(281) 17.288   (282) 60.1   (283) 54.7   (284) 20.7

(285) 20.88   (286) 10.9   (287) 19.223   (288) 39.9

DECIMALS

Basic Math Answer Keys

(289) 29   (290) 17.33   (291) 30.2   (292) 52.01

(293) 21.3   (294) 28.1   (295) 9.3   (296) 10.61

(297) 36.6   (298) 33.39   (299) 49.5   (300) 41.13

(301) 533.9   (302) 180.17   (303) 1.7   (304) 434.5

(305) 212.7   (306) 261.6   (307) 376.68   (308) 237.3

(309) 440.39   (310) 222.8   (311) 316   (312) 314.67

(313) 482.5   (314) 538.4   (315) 484.8   (316) 375.8

(317) 237.41   (318) 217.586   (319) 719.2   (320) 629.7

(321) 431.2   (322) 138.7   (323) 299.8   (324) 41.2

(325) 80.658   (326) 470.5   (327) 310.085   (328) 508.93

(329) 514.5   (330) 382.13   (331) 251   (332) 52.7

(333) 256.789   (334) 454.28   (335) 516.6   (336) 35.9

DECIMALS

Basic Math Answer Keys

(337) 585.4　　(338) 493.7　　(339) 309.901　　(340) 234.7

(341) 866.4　　(342) 402.67　　(343) 321.1　　(344) 307.95

(345) 463.1　　(346) 698.3　　(347) 353.3　　(348) 212.3

(349) 751.66　　(350) 528.3　　(351) 252.32　　(352) 111.3

(353) 322.1　　(354) 464.7　　(355) 372.5　　(356) 139.99

(357) 245.2　　(358) 466.4　　(359) 177.56　　(360) 455.44

(361) 808　　(362) 507.6　　(363) 84.3　　(364) 520.7

(365) 715.347　　(366) 298.5　　(367) 780.2　　(368) 660.77

(369) 85.4　　(370) 485.3　　(371) 193.307　　(372) 440.7

(373) 765.5　　(374) 561.51　　(375) 481.5　　(376) 197

(377) 478.7　　(378) 207.86　　(379) 17　　(380) 941.9

(381) 896.37　　(382) 572.4　　(383) 444.5　　(384) 714.2

DECIMALS

Basic Math Answer Keys

(385) 441 (386) 186.4 (387) 647.49 (388) 328

(389) 771.1 (390) 367 (391) 73.08 (392) 24.3

(393) 34.98 (394) 10.9424 (395) 147.4 (396) 45.14

(397) 76.575 (398) 68.3862 (399) 21.28 (400) 38.57

(401) 10.4 (402) 7.41 (403) 64.35 (404) 0.44

(405) 9.25452 (406) 139.974 (407) 33.88 (408) 51.87

(409) 81.74 (410) 62.31 (411) 2.278 (412) 18.55

(413) 35.77 (414) 4.84 (415) 131.445 (416) 91.5

(417) 50.4 (418) 2.5344 (419) 89.1 (420) 79.31

(421) 1.66 (422) 69.44 (423) 63.75 (424) 4.8

(425) 52.884 (426) 6.82 (427) 5.321 (428) 5.6

(429) 67.86 (430) 36.991 (431) 10.53 (432) 2.46

DECIMALS

Basic Math Answer Keys

(433) 33.6   (434) 141.9467   (435) 75.75   (436) 135.7268

(437) 10.01   (438) 5.92   (439) 94.44   (440) 99.28

(441) 1734.2236 (442) 491.41   (443) 899   (444) 191.77

(445) 40.32   (446) 1685.1   (447) 1544.48   (448) 853.74

(449) 1129.128   (450) 134.85   (451) 165.69   (452) 1099.62

(453) 271.93   (454) 487.5   (455) 1236.152   (456) 1037.85

(457) 1.55   (458) 32.58   (459) 40.66   (460) 303.8

(461) 24.65   (462) 344.97905   (463) 473.04   (464) 1297.35

(465) 1045.42   (466) 180.75   (467) 372.9   (468) 228.76

(469) 49.568   (470) 1479.62   (471) 1288.0436   (472) 30.97

(473) 873.775   (474) 330.78   (475) 1426.18   (476) 74

(477) 162.12   (478) 983.35   (479) 1635.5589   (480) 56.32

DECIMALS

Basic Math Answer Keys

(481) 202.401   (482) 1928.16   (483) 1158.74   (484) 14.8

(485) 1531.35   (486) 1944.45   (487) 1654.4   (488) 1326

(489) 36   (490) 270.9   (491) 4556   (492) 208.6948

(493) 994.9485   (494) 907.92   (495) 2965.92   (496) 3107.71584

(497) 84.52224   (498) 278.31756   (499) 1497.408   (500) 381.524

(501) 117.81   (502) 1037.184   (503) 299.0636   (504) 312.85152

(505) 1153.0064462 (506) 115.872   (507) 778.734   (508) 78.166

(509) 474.904   (510) 39.564   (511) 210.864   (512) 3.2725

(513) 16.764   (514) 635.44   (515) 1693.848   (516) 71.31593

(517) 561.66   (518) 242.2   (519) 1438.5584   (520) 87.971

(521) 1072.03   (522) 794.5344   (523) 636.23   (524) 81

(525) 1351.35   (526) 2497.824   (527) 74.646   (528) 256.704

DECIMALS

Basic Math Answer Keys

(529) 585.4983   (530) 78.526   (531) 154.9744   (532) 89.31

(533) 149.4   (534) 406.1   (535) 35.49   (536) 7.632

(537) 34.398   (538) 2879.37   (539) 336.6664   (540) 1135.232

(541) 0.6   (542) 3.5   (543) 38   (544) 1

(545) 1.42   (546) 2.8   (547) 5   (548) 1.75

(549) 0.2   (550) 0.2   (551) 0.9   (552) 29

(553) 5.2   (554) 1.75   (555) 1.75   (556) 0.4

(557) 30.5   (558) 7.5   (559) 1.44   (560) 2

(561) 11.75   (562) 1.05   (563) 0.5   (564) 71

(565) 149   (566) 0.15   (567) 3   (568) 2.98

(569) 5.56   (570) 58.5   (571) 5.2   (572) 23.25

(573) 0.5   (574) 3   (575) 1.5   (576) 15.25

DECIMALS

Basic Math Answer Keys

(577) 3.75   (578) 56   (579) 0.5   (580) 0.75

(581) 1.2   (582) 1.6   (583) 1.55   (584) 0.75

(585) 3   (586) 0.625   (587) 2.1   (588) 29.25

(589) 1.5   (590) 1.33   (591) 22   (592) 2

(593) 5.3   (594) 6.35   (595) 7.5   (596) 13

(597) 23.5   (598) 4.7   (599) 15   (600) 1.45

(601) 26   (602) 2.26   (603) 21   (604) 0.85

(605) 4   (606) 4.5   (607) 23.5   (608) 3.9

(609) 95   (610) 1.92   (611) 5.25   (612) 46.75

(613) 81.75   (614) 61   (615) 0.92   (616) 1.2

(617) 36.25   (618) 204.75   (619) 0.64   (620) 2.96

(621) 200   (622) 176.8   (623) 0.69   (624) 8.76

DECIMALS

Basic Math Answer Keys

(625) 2.8   (626) 1.3   (627) 125.5   (628) 0.55

(629) 6.25   (630) 0.548   (631) 17.65   (632) 8.37

(633) 7.24   (634) 135   (635) 2.54   (636) 14.43

(637) 26.5   (638) 0.46   (639) 0.7   (640) 50.5

(641) 150   (642) 7   (643) 22,270   (644) 9.0

(645) 6.2   (646) 5,009   (647) 1.2   (648) 993

(649) 79.0   (650) 2.0   (651) 250   (652) 72.9

(653) 866   (654) 8.8   (655) 10   (656) 29,807

(657) 41.0   (658) 521,370   (659) 609   (660) 3.5

(661) 476,789 (662) 7.3   (663) 212,490   (664) 4.1

(665) 87.0   (666) 7.8   (667) 4.30   (668) 3.12

(669) 3.9   (670) 9.0   (671) 0.18   (672) 0.1

DECIMALS

Basic Math Answer Keys

(673) 9.5   (674) 2.8   (675) 8.7   (676) 3.09

(677) 4.99   (678) 7.0   (679) 0.1   (680) 6.8

(681) 3.0   (682) 1.9   (683) 2.85   (684) 2.99

(685) 7.21   (686) 3.89   (687) 6.8   (688) 9.6

(689) 3.0   (690) 5.2   (691) 5.5219   (692) 9.0599

(693) 6.4101   (694) 4.438   (695) 4.1960   (696) 9.231

(697) 2.8434   (698) 5.4673   (699) 6.4617   (700) 7.1676

(701) 1.150   (702) 4.702   (703) 5.069   (704) 9.3683

(705) 9.6511   (706) 7.952   (707) 1.8880   (708) 3.8797

(709) 6.290   (710) 2.806   (711) 8.4920   (712) 9.734

(713) 5.244   (714) 3.3428   (715) 4.8974   (716) 4.64927

(717) 9.090144   (718) 8.546620   (719) 1.910284   (720) 8.56121

DECIMALS

Basic Math Answer Keys

(721) 9.78126   (722) 8.242774   (723) 3.624541   (724) 1.24443

(725) 2.95569   (726) 7.98539   (727) 4.19452   (728) 5.42252

(729) 9.111860   (730) 4.587800   (731) 5.15749   (732) 2.18049

(733) 6.55566   (734) 4.208619   (735) 9.10869   (736) 1.095030

(737) 9.900815   (738) 8.56555   (739) 3.43669   (740) 1.25950

(741) four tenths   (742) nine tenths   (743) two tenths   (744) three tenths

(745) five tenths   (746) forty-nine hundredths   (747) fifty-two hundredths

(748) seven hundredths   (749) six hundredths

(750) fifty-five hundredths   (751) eight hundredths

(752) ninety-six hundredths   (753) thirty-two hundredths

(754) three hundredths   (755) seventy-five hundredths

(756) six hundred twenty-one thousandths   (757) two hundred sixteen thousandths

DECIMALS

Basic Math Answer Keys

(758) three hundred one thousandths

(759) five hundred fifty-five thousandths

(760) twenty-seven thousandths

(761) eight thousandths

(762) seven hundred fifty-six thousandths

(763) eighty-six thousandths

(764) one thousandth

(765) five hundred ninety-seven thousandths

(766) nine hundred eight ten-thousandths

(767) five hundred seventy-four ten-thousandths

(768) four thousand, seven hundred forty-two ten-thousandths

(769) eight hundred forty-eight ten-thousandths

(770) four hundred two ten-thousandths

(771) one ten-thousandth

(772) seven hundred seven ten-thousandths

(773) sixty-four ten-thousandths

(774) four thousand, two ten-thousandths

(775) four thousand, twenty-two ten-thousandths

(776) nine thousand, seven hundred ninety-three hundred-thousandths

(777) six thousand, nine hundred seventy-five hundred-thousandths

(778) one thousand, two hundred eighty-two hundred-thousandths

(779) four thousand, five hundred-thousandths

(780) three thousand, five hundred nine hundred-thousandths

(781) fifteen thousand, three hundred thirty-eight hundred-thousandths

DECILMALS

Basic Math Answer Keys

(782) eighty-seven thousand, seventy-eight hundred-thousandths

(783) twenty thousand, eight hundred nineteen hundred-thousandths

(784) seventy-seven hundred-thousandths

(785) twenty-seven thousand, three hundred four hundred-thousandths

(786) three hundred nine thousand, five hundred three millionths

(787) twenty-five thousand, eight millionths

(788) seven hundred thirty thousand, three hundred four millionths

(789) four hundred ninety-three millionths

(790) two hundred nineteen thousand, three millionths

(791) one millionth

(792) five hundred eight millionths

(793) one hundred seven millionths

(794) nine hundred thousand, twenty-nine millionths

(795) twenty-four thousand, nine hundred eight millionths

(796) six ten-millionths

(797) sixty thousand, forty-eight ten-millionths

(798) seven million, twenty thousand, seven ten-millionths

(799) eight million, six hundred ninety thousand, seven ten-millionths

(800) forty-three thousand, four hundred fifty-seven ten-millionths

(801) thirty-one ten-millionths

(802) eight ten-millionths

(803) five hundred seventy thousand, two hundred eighty-nine ten-millionths

(804) nine hundred four thousand, five ten-millionths

DECIMALS

Basic Math Answer Keys

(805) four million, five hundred eighteen thousand, seven hundred seventy-three ten-millionths

(806) thirty million, three hundred sixty thousand, two hundred twenty-four hundred-millionths

(807) nine million, three hundred fifteen thousand, six hundred four hundred-millionths

(808) twenty-six million, eight hundred eight hundred-millionths

(809) seventy million, ninety thousand, one hundred six hundred-millionths

(810) sixty million, eight thousand, thirty-eight hundred-millionths

(811) sixty-two million, three hundred thousand, three hundred-millionths

(812) fifty million, forty-eight hundred-millionths

(813) seventy million, six thousand, twenty-five hundred-millionths

(814) forty-three million, one hundred ninety thousand, nine hundred-millionths

(815) ten million, twenty-two thousand, three hundred thirteen hundred-millionths

(816) six hundred three million, eight thousand, sixty-four billionths

DECIMALS

(817) ninety-five million, two hundred thirty-two thousand, eighty-one billionths

(818) one hundred sixty-three million, three thousand, one billionths

(819) six hundred seventy-six million, one hundred ten thousand, ninety-four billionths

(820) eight hundred million, one hundred fifty thousand, two hundred one billionths

(821) four hundred eighty-six million, six hundred seven thousand, one hundred six billionths

(822) two hundred million, one hundred three thousand, eight billionths

(823) one hundred six million, eight thousand, eight hundred three billionths

(824) five hundred million, eight hundred six thousand, eighty-six billionths

(825) three hundred sixty-one million, six hundred five thousand, sixty-three billionths

(826) 0.2

(827) 0.4

DECIMALS

Basic Math Answer Keys

(828) 0.3   (829) 0.8   (830) 0.7   (831) 210.64

(832) 860.27   (833) 500.04   (834) 550.98   (835) 700.62

(836) 150.73   (837) 197.06   (838) 161.98   (839) 300.02

(840) 107.09   (841) 105.01   (842) 130.33   (843) 305.35

(844) 706.88   (845) 610.76   (846) 309.48   (847) 406.81

(848) 590.05   (849) 700.03   (850) 691.08   (851) 607.009

(852) 908.084   (853) 154.708   (854) 970.092   (855) 550.058

(856) 270.071   (857) 407.876   (858) 802.067   (859) 648.606

(860) 254.408   (861) 600.599   (862) 657.096   (863) 709.568

(864) 600.349   (865) 330.508   (866) 754.002   (867) 967.506

(868) 490.027   (869) 969.051   (870) 800.805

(871)   79,380,046.9153   (872)   964,708,990.4514

DECIMALS

Basic Math Answer Keys

(873) 845,540,305.1004   (874) 4,040,103.0009

(875) 2,843,187.6155   (876) 2,856,650.1303

(877) 4,600,083.5383   (878) 580,068,405.4482

(879) 89,101,306.4007   (880) 89,340,833.8098

(881) 103,006,800.8905   (882) 5,406,300.3004

(883) 9,000,020.0975   (884) 23,701,020.1159

(885) 70,081,009.6013   (886) 104,303,160.5033

(887) 8,609,080.0907   (888) 420,009,030.8913

(889) 3,700,090.0739   (890) 8,500,013.7006

(891) 16,959,200.09425   (892) 100,186,870.00002

(893) 144,124,650.06092   (894) 60,495,500.80091

(895) 503,000,950.03866   (896) 99,750,600.01217

DECIMALS

Basic Math Answer Keys

(897) 605,877,002.79708

(898) 608,107,006.41106

(899) 10,000,025.00479

(900) 42,050,088.06901

(901) 4,400,100.04309

(902) 20,302,002.06842

(903) 900,270,806.48109

(904) 26,649,200.91031

(905) 58,420,050.70723

(906) 200,108,620.05008

(907) 5,705,924.00503

(908) 94,666,100.38002

(909) 928,331,000.00082

(910) 674,005,004.40305

(911) 580,180.3080663

(912) 8,043.3000701

(913) 897,000.4063007

(914) 3,013.0725076

(915) 980,320.1220095

(916) 7,040.7057008

(917) 300,020.7500209

(918) 90,084.0098049

(919) 20,681.0708003

(920) 20,000.3000627

DECIMALS

Basic Math Answer Keys

(921) 751,840.8186202

(922) 69,300.6204453

(923) 76,004.0870017

(924) 662,077.6400072

(925) 7,600.0200004

(926) 8,006.8000951

(927) 18,040.6000307

(928) 9,004.0921903

(929) 608,924.0030083

(930) 3,090.2891487

(931) 7,290,053,150.04063701

(932) 65,208,304,980.37007509

(933) 3,085,080,791.20960095

(934) 70,004,204,603.60043024

(935) 89,075,008,096.04600006

(936) 141,000,000,310.00340047

(937) 4,320,001,085.08079178

(938) 230,001,309,653.70508705

(939) 60,320,209,001.00201005

(940) 75,904,570,540.13022007

(941) 9,310,000,806.99079004

(942) 7,109,700,010.61100095

(943) 381,271,740,100.00700006

(944) 40,030,998,283.77009041

DECIMALS

Basic Math Answer Keys

(945) 5,097,003,760.08850008

(946) 90,720,830,080.10000646

(947) 56,001,085,080.79179607

(948) 91,600,070,500.44008109

(949) 180,010,460,699.80300403

(950) 70,903,090,490.06830382

(951) tenths

(952) hundredths

(953) tenths

(954) tenths

(955) ones

(956) ones

(957) ones

(958) tenths

(959) tenths

(960) hundredths

(961) ones

(962) ones

(963) hundredths

(964) tenths

(965) tenths

(966) tenths

(967) hundredths

(968) ones

(969) ones

(970) tenths

(971) tenths

(972) ones

(973) ones

(974) hundredths

(975) hundredths

(976) hundredths

(977) ones

(978) tenths

(979) ones

(980) tenths

(981) ones

(982) ones

(983) ones

(984) ones

DECIMALS

Basic Math Answer Keys

(985) ones   (986) hundredths   (987) hundredths   (988) ones

(989) ones   (990) hundredths   (991) hundredths   (992) tenths

(993) hundredths   (994) hundredths   (995) tenths   (996) tenths

(997) hundredths   (998) ones   (999) tenths   (1000) ones

(1001) hundredths   (1002) thousandths   (1003) hundredths   (1004) thousandths

(1005) hundredths   (1006) hundredths   (1007) thousandths   (1008) hundredths

(1009) thousandths   (1010) thousandths   (1011) thousandths   (1012) hundredths

(1013) thousandths   (1014) hundredths   (1015) hundredths   (1016) thousandths

(1017) thousandths   (1018) thousandths   (1019) thousandths   (1020) hundredths

(1021) hundredths   (1022) hundredths   (1023) thousandths   (1024) hundredths

(1025) thousandths   (1026) hundredths   (1027) thousandths   (1028) hundredths

(1029) thousandths   (1030) hundredths   (1031) hundredths   (1032) thousandths

DECIMALS

Basic Math Answer Keys

(1033) hundredths (1034) thousandths (1035) hundredths (1036) hundredths

(1037) hundredths (1038) thousandths (1039) hundredths (1040) hundredths

(1041) thousandths (1042) hundredths (1043) thousandths (1044) hundredths

(1045) thousandths (1046) thousandths (1047) hundredths (1048) hundredths

(1049) hundredths (1050) thousandths (1051) hundred-thousandths

(1052) ten-thousandths (1053) ten-thousandths

(1054) ten-thousandths (1055) ten-thousandths

(1056) ten-thousandths (1057) ten-thousandths

(1058) ten-thousandths (1059) millionths

(1060) ten-thousandths (1061) hundred-thousandths

(1062) ten-thousandths (1063) ten-thousandths

(1064) millionths (1065) hundred-thousandths

DECIMALS

(1066) millionths

(1067) ten-thousandths

(1068) ten-thousandths

(1069) millionths

(1070) ten-thousandths

(1071) hundred-thousandths

(1072) ten-thousandths

(1073) ten-thousandths

(1074) hundred-thousandths

(1075) millionths

(1076) ten-thousandths

(1077) millionths

(1078) ten-thousandths

(1079) ten-thousandths

(1080) hundred-thousandths

(1081) ten-thousandths

(1082) ten-thousandths

(1083) hundred-thousandths

(1084) millionths

(1085) ten-thousandths

(1086) ten-thousandths

(1087) millionths

(1088) hundred-thousandths

(1089) ten-thousandths

DECIMALS

Basic Math Answer Keys

(1090) ten-thousandths

(1091) millionths

(1092) ten-thousandths

(1093) millionths

(1094) ten-thousandths

(1095) ten-thousandths

(1096) hundred-thousandths

(1097) ten-thousandths

(1098) millionths

(1099) hundred-thousandths

(1100) hundred-thousandths

(1101) 17.3

(1102) 15

(1103) 700

(1104) 53.7

(1105) 1.0

(1106) 4

(1107) 9.30

(1108) 7.68

(1109) 2.46

(1110) 30.39

(1111) 1

(1112) 0.9

(1113) 91.0

(1114) 5.1

(1115) 9

(1116) 71,999

(1117) 46.94

(1118) 393

(1119) 5.7

(1120) 46.60

(1121) 1,329

(1122) 9

(1123) 4.80

(1124) 5.08

(1125) 9

DECIMALS

Basic Math Answer Keys

(1126) 217,595   (1127) 26.3   (1128) 6.0   (1129) 400,125

(1130) 37.4   (1131) 5.9   (1132) 3   (1133) 1.6

(1134) 7.65   (1135) 671   (1136) 70.0   (1137) 39

(1138) 69,389   (1139) 5.40   (1140) 6.80   (1141) 7.1

(1142) 440   (1143) 5.7   (1144) 7.9   (1145) 8.02

(1146) 1.69   (1147) 5.9   (1148) 1.57   (1149) 9.4

(1150) 9.3   (1151) 0.30713   (1152) 5.066015   (1153) 5.6552

(1154) 8.69969   (1155) 3.953220   (1156) 2.408   (1157) 4.338109

(1158) 4.700   (1159) 9.490   (1160) 0.6599   (1161) 1.039

(1162) 2.920609   (1163) 5.5697   (1164) 2.3888   (1165) 4.57737

(1166) 5.327669   (1167) 4.0500   (1168) 8.10023   (1169) 8.730

(1170) 1.469   (1171) 9.23794   (1172) 6.651   (1173) 9.262

DECIMALS

Basic Math Answer Keys

(1174) 3.135634	(1175) 3.074	(1176) 4.99958	(1177) 3.670

(1178) 6.7784	(1179) 1.587120	(1180) 7.6298	(1181) 2.19289

(1182) 7.9150	(1183) 3.5665	(1184) 6.7754	(1185) 6.951205

(1186) 5.276167	(1187) 6.3592	(1188) 1.807217	(1189) 1.870

(1190) 1.81003	(1191) 3.1664	(1192) 3.012530	(1193) 4.002

(1194) 9.641509	(1195) 6.657	(1196) 1.41855	1197) 3.195520

(1198) 5.2190	(1199) 5.98165	(1200) 9.1413